CW00686867

SOAPDISH EDITIONS

Organic Bath

CREATING A NATURAL, HEALTHY HAVEN

A WATERPROOF BOOK

Text by Kyle Roderick
Illustrations by Becky Heavner

MELCHER MEDIA

Published in the UK by
Dorling Kindersley Limited
80 Strand, London WC2R 0RL
www.dk.com
in association with

124 West 13th Street
New York, NY 10011
www.melcher.com

Publisher: Charles Melcher
Associate Publisher: Bonnie Eldon
Editor in Chief: Duncan Bock
Project Editor: Betty Wong
Assistant Editor: Lindsey Stanberry
Production Director: Andrea Hirsh

Design by 3+Co.

Special thanks to: David E. Brown, Heidi Ernst, Lauren Nathan, Lia Ronnen, Holly Rothman, Alex Tart, Shoshana Thaler, and Megan Worman.

09 08 07   10 9 8 7 6 5 4 3 2 1

Printed in China
A CIP catalogue record for this book is available from the British Library.
ISBN 978-1-59591-032-5

First Edition

Table of Contents

Introduction: A Healing Space 8

Creating a Natural Bath 14

Nourishing the Senses 56

Organic Body Care 86

INTRODUCTION

A Healing Space

You may be familiar with the saying "All life began in the sea." And you may have noticed that regardless of how tired or stressed you are, bathing always renews you and improves your mood. But have you considered bringing life to your bathroom by making it a healing spa space?

With a few simple changes, you can transform your bath into a true haven for resting and resetting body, heart, mind, and spirit.

The best way to create an idyllic retreat is by filling your bathroom with organic products and earth-friendly materials. Most common bath and body products, as well as conventional bathroom cleaners and building materials, contain irritating or toxic chemicals that pose health risks to you and our planet.

Because lotions and potions are absorbed through the skin, beauty products made from organically grown botanicals and sea minerals are safest and best for our bodies, skin, and hair. Rich in natural ingredients and free of toxins, they also make you look and feel good. What's more, they provide the priceless health and beauty benefit called peace of mind.

This peace permeates your being, thanks to the aromatic properties of organic plants, fruits, and flowers, which can positively influence physiology and moods. Using organic products for staying clean and serene can help slow brain waves, lower blood pressure, enhance visual tracking, relieve cold and flu symptoms, prime you for falling asleep, or get you in the mood for love.

As you are about to learn in detail, creating an eco-friendly bath space (and living a greener lifestyle) doesn't mean compromising on beauty or luxury. Moreover, you'll find that many of the suggestions here are inexpensive to follow, while providing multiple bonuses. You'll find ideas for creating, stocking, and maintaining a natural bathroom, as well as simple recipes for making your own skin- and hair-care products.

Whenever you nourish all five senses in the bath or shower, the bathing ritual becomes a treasured time.

Enjoy.

CHAPTER 1

Creating a Natural Bath

The first step toward designing your oasis involves walking into your bathroom, looking around, and taking an inventory. Unless you are planning to remodel, you can just make several simple and affordable changes to create a more soothing environment. Think: out with the old, synthetically made things, and in with the organic or recycled items.

Here are some eco-friendly materials to look for:

- Water- or milk-based paints and nontoxic, biodegradable stains

- Recycled or salvaged shelving, furniture, mirrors, light fixtures, and fabrics
- Organic cotton, hemp, bamboo, linen, soy, and other sustainable fabrics
- Jute, sisal, and cork flooring
- Sustainably harvested woods and bamboo

bh

FURNITURE AND ACCESSORIES

....

Take a good, hard look at your clutter quotient. What can you throw out right now? Clean out cupboards; toss old cosmetics and items that you haven't used for a while. Clear away stray boxes and bottles from under your sink or elsewhere.

Many apartments or homes have plastic medicine cabinets; perhaps you've got one as well. Consider replacing it with a wooden one, but if you can't reuse the plastic cabinet or give it away, keeping it could be preferable to having it end up in a landfill. Should you keep yours, you can always paint it a more restful color.

Plastic or vinyl shelves and hampers are next to go. Perhaps these can be used in your garage or laundry room. If not, donate them to a secondhand store. (If you call and explain what you are getting rid of, many secondhand stores will even come and cart away your goods, depending on volume.)

When searching for replacements (or simply more adequate storage space), spend time considering what kind of objects and materials best suit your needs:

* Unstained woods or basketry can help you commune with nature.
* Shelves or towel racks painted in soothing colors can help your body and mind calm down and rest.
* Recycled or antique furnishings can create a vintage haven.

Choosing bath towels, washcloths, robes, bath mats, and shower curtains made from organic fabrics is a smart decision. Organically grown plant fibers are hardier, softer, and literally denser than conventionally grown plant fibers. They are also stronger and smoother, having never been chemically fertilized or doused with pesticides and herbicides (many of which contain known carcinogens), and they will last at least twice as long as fibers in conventional cotton towels.

Some buying tips:

- Check national health-food chains or the Internet for organic cotton or bamboo towels.
- Instead of vinyl or plastic shower curtains, try those made of linen, organic cotton, or hemp, which has antimicrobial properties.

- Instead of plastic soap dispensers and dishes, try those made from natural materials such as seashells, bamboo, wood, or stone.
- Last but not least: Plastic and vinyl toilet seats are standard accessories in twenty-first-century bathrooms. Instead, choose an eco-friendly bamboo or wood one—they are more aesthetically appealing and often more comfortable than synthetic styles.

FLOORING

Perhaps you're thinking about replacing your bathroom floor. But even if your bathroom has linoleum or other synthetic flooring that you can't get rid of now, you can still easily ground your space with functional organic elements. Your choices will be determined by budget, creativity, and time constraints. Here are some ideas to get you started:

- Cover existing flooring with rugs or woven mats made from natural fibers such as jute or sisal, and use organic cotton or wooden bath mats.
- Check out mildew-resistant cork flooring, although know that it requires a layer of preparatory surface below it.
- Remove existing paint or varnish from wood floors and stain them with a biodegradable, plant-based oil such as linseed oil. Take care to seal in the stain with a nature-based floor wax.

- Remember to use biodegradable, water-soluble stains and paints if you are painting your floor.
- Replace old, tired tile flooring with natural stone or recycled glass in a delightful design. Admittedly, this is a time-consuming option, but creative tile colors can fill your bathroom with beauty, serenity, and one-of-a-kind charm. If you're the do-it-yourself type, you can even lay the tile on your own.

WALLS AND CEILINGS

·······

Could your bathing sanctuary use a new coat of paint? If so, do a little research to find a source that carries eco-friendly, nontoxic, biodegradable paints. Check out color swatches, and contemplate what effect you want to create in your water world.

For the most healing and soothing environment, think rose. Why? Because pink dramatically relaxes us more than any other color. It lowers blood pressure and heart rate, thus fostering peaceful emotions. Other colors that foster serenity include robin's-egg blue, spring green, and yellow—which is known to help improve concentration.

 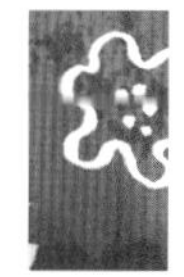 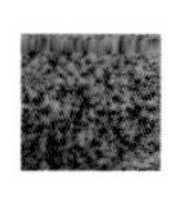   

LIGHTING AND MIRRORS

....

Saving the planet extends even to lightbulbs. While energy-efficient compact-fluorescent bulbs might cost more initially, they last far longer than other bulbs and significantly cut electricity costs in the long run.

- Generate spa-mood lighting by installing a dimmer switch.
- Create a zen atmosphere with a white paper-lantern fixture.

- If retro is your thing, shop for antique light fixtures.
- Add a little drama with crystal or glass chandeliers, which radiate vibrant illumination and can produce striking effects if wired to a dimmer switch. Chandeliers are especially welcome in rainy climates and windowless bathrooms. Strategically placed wall mirrors will reflect light, allowing you to bathe by a glamorous glow.

Mirrors can also create the illusion of more space, which in turn can relax you. Look for mirrors with reclaimed metal frames, or enhance wooden frames with organic elements. You can easily add seashells, driftwood, or tree bark with a glue gun.

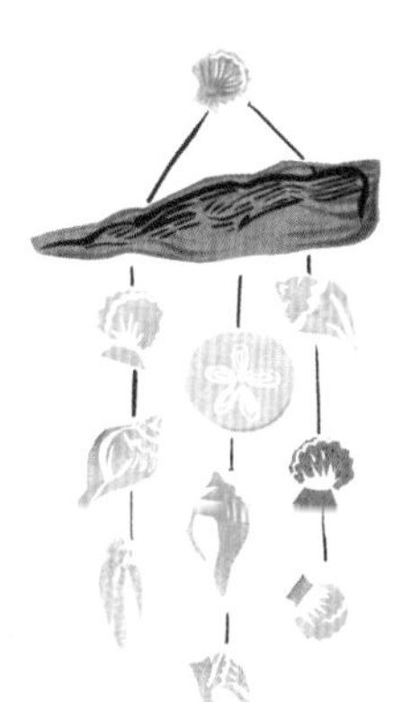

WATERWISE FILTERS, TAPS, TOILETS, TUBS, AND SHOWERS

....

Life on our planet exists thanks to our oceans, lakes, rivers, aquifers, and the rains that fall from above. But because of industrialization, our rain and earth waters contain thousands of chemicals, many of whose long-term health effects are unknown. Each year hundreds of new chemicals are invented, thus entering the water table and our bodies. More than seven hundred toxic chemicals, including chlorine, have been routinely identified in the world's drinking and bathing water. We absorb these through the skin and inhale them while we bathe, brush our teeth, shampoo the dog, or hand-wash clothes. One way to avoid the chemicals is to purify your sink taps, bath taps, and showerheads with filters.

Given that efficient water usage is integral to an organic bathroom, consider installing a low-flow showerhead, and

either a low-flush toilet or a dual-flush toilet (dual-flush toilets enable users to save 25% more water than low-flush models). Do some consumer research online to find the best models for your home.

When it comes to saving hot water, cast-iron bathtubs retain heat longer than those made of synthetic materials. Vintage cast-iron tubs are also generally a few inches deeper, which means your body can more easily sink in for a long, comfortable soak.

AIR QUALITY AND HEATING

....

Breathing is your body's primary critical function, so your organic sanctuary should have air that's as pure and healthy as possible. Common volatile organic compounds (known as VOCs) are chemicals used to manufacture many interior furnishings, cleaners, and textiles. They can float around your house and into the bathroom, irritating lungs, causing sinus conditions, and triggering skin flare-ups. But here is a breath of fresh air: A NASA study has proven that twelve easily sourced houseplants can help remove VOCs from indoor spaces.

The top houseplants for removing concentrations of formaldehyde in the air are the green spider plant, bamboo palm, dracaena "Janet Craig," mother-in-law's tongue, dracaena marginata, peace lily, and golden pathos. Potted chrysanthemums, English ivy, and gerbera daisies also help cleanse and beautify indoor oases with

living, breathing greenery. Note: two to three plants in eight- or ten-inch pots will help purify the air in the typical-size bathroom.

Remember, when possible, to keep windows open while showering and bathing to help steam and humidity escape.

If the temperature inside your bathroom feels arctic, judicious use of an energy-efficient space heater may be the answer.

And if you live in a damp waterside area or rainy region, installing electrically heated towel racks makes excellent sense. Besides preventing towels from getting mildewed, these help wrap your body in much-needed warmth after a bath. Stainless-steel styles are better than plastic because they won't crack or split.

EARTH-FRIENDLY CLEANERS

....

A few years back, a study on the dangers of common household cleaning products made front-page news. The investigators found that each and every house stocked floor and sink cleaners that variously contained hormone-disrupting, toxic, or cancer-causing agents.

Before the chemical industrial complex started marketing these toxic, non-biodegradable cleaning

products a few generations back, most tub, sink, and toilet cleaning was done with simple, effective, and low-cost ingredients: white vinegar, lemon juice, baking soda, boiling water, and earth-friendly scouring powders such as Bon Ami Polishing Cleanser. And guess what? These are still all you need to effectively clean your bathroom today.

- White vinegar (full strength) cleans everything in the bathroom, including toilets and the tub, and leaves a fresh fragrance. It is especially good for removing soap

and toothpaste from mirrors and making them sparkle.

- Defog mirrors with Bon Ami (made from calcite and feldspar mineral abrasives and biodegradable detergent) on a wet sponge.
- To clean toilet bowls, pour in boiling hot water, then sprinkle in some Bon Ami or baking soda. Any stains will quickly be scrubbed away.
- Brighten up tile floors with a paste of baking soda, warm water, and a few drops of lemon juice.

It's also worth noting that many supermarkets, department stores, health-food stores, and online retailers sell nontoxic, biodegradable cleaners that come in convenient spray bottles.

ORGANIC BATHING TOOLS

....

Fortunately, tools for the organic bath are inexpensive, effective, and easy to come by.

- If you don't already own one, by all means get a natural sea sponge to clean your body. They're softer, more absorbent, and much more decorative than synthetic sponges.

bh

- You'll also need a wooden-and-natural-fiber nail brush for prettifying fingernails and toenails.
- Get a loofah mitt for exfoliation.
- Buy a wooden-and-natural-fiber body brush. It is essential for skin brushing.
- For after-bath sprucing, you'll also need a natural pumice stone for removing heel calluses and polishing the underside of your feet.

CHAPTER 2

Nourishing the Senses

Now that you've designed and stocked your organic bath, make the most of your sanctuary by feeding all five senses.

Numerous aromas, sounds, sights, tastes, and things to touch will quickly promote mind-body relaxation and improve moods. Try the following quick and easy products and strategies for enriching your senses and refreshing every aspect of your being.

SCENT

Essential Oils Organic essential oils are fragrant tools for enhancing the health benefits of your bath. An essential oil consists of a highly concentrated plant extract, such as rose. It is often mixed with a plant-based carrier oil. Jojoba oil's chemical composition is close to human sebum, the skin oil that our bodies naturally produce, and it is the most popular carrier for essential oils.

As you may already know, certain scents, such as lavender, have been scientifically proven to slow brain waves, lower blood pressure, relax muscles, and promote the onset of sleep.

Uplifting ones—such as rosemary, citrus, and cypress—are useful when you need to wake up, clear your mind, energize your body, and get moving.

Eucalyptus, thyme, clary sage, and Scots pine essential oils can help relieve congestion.

athroom with the healing aromas of
ial oils, try one or more of the following.
for maximum benefits.
atherapy diffuser to waft essential oil
air before, during, and after your bath.
v drops of essential oil into your
scent your bath and help improve
hysical symptoms.
essential-oil-scented candles as you bathe.

Melissa, chamomile, ylang-ylang, and jasmine all help slow brain waves, elicit relaxation, and ease stress. Research indicates that jasmine activates the limbic system in the brain and stimulates the rod-shaped cells in the retina, on which we rely heavily for night vision. Because jasmine enhances visual tracking and night vision, you may want to use it in your bath ritual when preparing for a night of driving, dancing, or working overtime.

Mixed floral scents of various types have also been

found to enhance hand an
coordination. Examples ir

- Romantic and sweet: y
jasmine, and rose
- Clean: lavender, iris,
and narcissus
- Tropical: gardenia, pi
and plumeria

To fill your
organic essen
Breathe deepl

- Use an arom
through the
- Sprinkle a fe
bathwater to
moods and p
- Burn organic

Incense and Potpourri Dried sage contains powerful essential oils and has long been burned by Native Americans for purification purposes. Sage incense clears the air and calms the mind with fragrant herbal smoke; burn a little before you get in the bath to freshen the room. (Many health-food stores stock dried sage bundles; it is quite affordable and lasts a long time.)

Potpourri is easy and inexpensive to make yourself. Simply mix together some dried flower buds and/or hard spices (such as cinnamon and cloves), then place in a bowl on a bathroom shelf.

SOUND

....

If there's a window in your bathroom, hang wind chimes to make pleasing sounds in the breeze as you shower or bathe.

Filling the room with soothing music or nature sounds adds yet another healing dimension to your space. (Just be sure your CD player sits a safe distance from the water.)

Nature sounds and music that cycle at sixty beats per minute or less elicit what is called the relaxation response. This is the medical condition during which heart rate slows, blood pressure decreases, and brain waves shift into theta—the creative "flow" state that is often only accessed during the dream stage of sleep, lovemaking, artistic creation, or physical exercise.

Examples of sounds and music that cycle at sixty beats per minute or less include:

- Slow Bach and other classical music
- Ocean waves
- Certain country music and blues
- Tibetan chanting
- Native American chanting
- Gregorian chanting
- Cool jazz
- Kirtan
- Traditional degung for flute and percussion

LIQUID SOUND HEALING

....

Lucky guests at the German hydro spas Toskana Therme Bad Sulza and Toskana Therme Bad Schandau bathe in the healing powers of sound, water, light, touch, and scent.

At these retreats, pools are wired with underwater speakers to create aquatic surround sound, or Liquid Sound, a system designed by Micky Remann. The pools are heated and filled with purified water and Epsom salts so that bathers float weightlessly while music and nature sounds travel through their bodies. Soothing lights and colors are projected onto the ceiling of the dome enclosing the pool, and fragrant essential oils are periodically released into the air. Germany's health-care system provides medical coverage for this treatment, said to induce deep relaxation and aid sleep disorders.

Try these suggestions to easily make your bathroom a healing liquid-sound temple:

* To renew energies and commune with the life force: play recordings of rainwater, waterfalls, and thunderstorms. Also try rainforest environments or ocean waves rolling to shore.
* To enhance physical healing: rolling ocean waves will also help slow your heart rate, brain waves, and nervous system, thereby comforting your body and helping you heal.
* To nurse a broken heart: fill your bath space with

dawn birdsongs, dolphin and whale songs, female Gregorian chanting, Sanskrit prayers, or your world music of choice.

* For a warm chill-out: tune in to slow jazz, chamber music, acoustic world, bamboo flute, or Native American flute recordings.
* To promote serenity and support contemplation: try your world music of choice or traditional Balinese (degung) acoustic melodies.

More good news: Several peer-reviewed medical research studies indicate that certain sounds and music can help improve concentration, increase productivity, and speed wound recovery.

SIGHT

....

There's nothing quite like basking in a luscious bath in the glow of candlelight. Scented candles make the bath even more enjoyable.

Most of the world's candles are made primarily from paraffin wax and animal fats. As a byproduct of gasoline refining, paraffin is neither biodegradable nor water-soluble. Soy or beeswax candles with 100% cotton wicks and organic essential oils will burn cleaner, cooler, and longer than paraffin candles, and without blackening the walls.

TASTE

....

During hot weather: Fill a pitcher with spring or sparkling mineral water; add thin cucumber and lemon slices. Drink a few glasses while soaking in the tub.

During winter weather: Brew a pot of cinnamon spice tea and sip slowly while bathing.

For a zen bath: Drink Japanese green tea.

For dispelling colds and flus: Grate a claw of ginger into a teapot and pour boiling water over it. Let steep for ten minutes, then drink while in the bath.

TOUCH

....

Any bath can be a sensual refuge, but you can make yours a super-tactile experience through the enlightened use of skin pleasers. For example, try a soothing bath oil that instantly smoothes your body and feet; naturally soft sea sponges that clean with a caress; and natural pumice

stones that remove dead and dry skin from feet and ankles. When you leave the tub or shower, wrap your body and head in thick, plush towels for warmth and comfort.

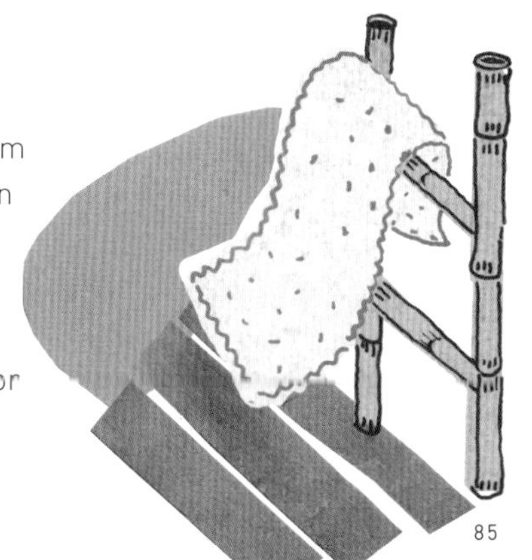

CHAPTER 3

Organic Body Care

BENEFICIAL PRE-BATH RITUALS

....

Simple rituals prepare your mind and body to derive maximum comfort from your bath's healing waters. Let the weather—along with your skin type, mood, health conditions, and time constraints—determine what you do during your pre-bath spa time.

You may, for example, apply a face and neck mask or warm yourself up with a full-on body scrub. Then again,

you might give yourself
a stimulating body-brush
treatment.

Skin Brushing Skin brushing has been practiced for millennia in many cultures. This exfoliating self-massage prepares you inside and out for showering or relaxing in the bath.

A brisk brushing with a wooden, natural-bristle body

brush, or a loofah, stimulates blood circulation and literally warms your heart—and the rest of your body. The enhanced circulation makes you feel more alert yet relaxed.

Start at your ankles and rub the brush or loofah in firm, circular strokes in the direction of your heart. Along with enhancing lymphatic-system functioning and drainage, skin brushing exfoliates and tones underneath skin to help reduce the appearance of cellulite.

Skin-Savvy Scrubs and Masks No discussion of pre-bath or pre-shower pampering is complete without detailing the various benefits of scrubs and masks. Here's a quick overview, organized by complexion type:

For normal and combination skin:

- Clay masks for pore cleansing, exfoliating, and smoothing You can find dried clay and/or clay mask creams at most health-food stores and online organic skin-care retailers. Because you must prepare it yourself with water, dried clay is usually less expensive than ready-to-use masks. Apply clay to your body before showering or while the tub is filling with water. Shower off before soaking, or wear the mask in the tub and remove it with bathwater.

- Sugar and essential-oil scrubs for pore cleansing, exfoliating, and moisturizing

  This kind of scrub deep cleans and infuses skin with oils that keep it from drying out in the bath. Many retailers also carry high-quality handmade scrubs.

The most cost-effective option is to make your own: You'll find both organic sugar and essential oils at the health-food store or online. The ratio of sugar to oils depends on how much scrub you want to prepare (it's best to make yours fresh every time). Fill a bowl with enough sugar to cover your face or body. Sprinkle the essential oil over the sugar until it forms a paste that you can easily apply with your fingers. Massage into skin and inhale the restful aromas.

For dry and sensitive skin:

- Organic oils

For thousands of years, massaging dry and sensitive skin with oil has been a vital pre-bath ritual. While the ancient Romans bathed in olive oil before soaking in hot springs, Polynesians applied coconut or tamanu oils. In the Indian healing tradition of Ayurveda, sesame oil is massaged into skin to calm and fortify the largest organ of the body. As oil from the jojoba plant is most similar in composition to your skin's natural oils, you may prefer to use this.

While all of the above oils nourish skin, organic olive oil offers the best value for your money, followed by sesame. To find organic coconut or tamanu oils, a health-food store is your best bet. Hint: Warming the oil before use makes application more luxurious while also enhancing absorption.

For oily skin:

- Clay masks for face and body
  Apply dried clay from a health-food store mixed with water for head-to-toe pore cleansing. Shower off or wear in the bath for further purification.
- Scrubs of sea salt and organic essential oils
  Rich in minerals, a sea salt scrub helps polish off blackheads, oil, and dead skin cells. The skin-drying properties of sea salt are useful in hot, humid weather.

Sea salt scrubs containing organic essential oils are easy to find at health-food and beauty-supply stores. You can also easily whip up your own by blending sea salt from a health-food store with jojoba or olive oil. Depending on whether you are blending a scrub for your face, chest, or entire body, fill a bowl with enough sea salt to cover the areas you wish to treat. Add an aromatic essential oil, a half-vial at a time, to create a paste that you can easily apply with your fingers.

Breathe in the healing scents and scrub gently with a soft washcloth. Rinse before leaving the tub or shower to remove all salt particles.

* Organic exfoliating scrubs

  Many organic skin-care manufacturers make scrubs for oily skin that contain oatmeal and ground walnut shells or apricot pits.

HOW TO IDENTIFY ORGANIC BATH PRODUCTS

As you probably already know, differentiating between earth-friendly body-care items and pseudo-organic ones can be a challenge. Labels and advertising are often misleading. Here are some tips from skin-care product formulator and spa consultant Jenefer Palmer:

* Read labels before buying. Look for products identified as "certified organic." This certification means that the Organic Trade Association

and other industry boards that confer organic status on consumer products have evaluated the product.

- Be advised that certain widely used ingredients are instant tip-offs that a product is neither organic nor biodegradable:

- Petroleum byproducts such as mineral oil, methylparaben, propylparaben, or other parabens
- Chemical foaming agents such as sodium lauryl sulfate and sodium laureth sulfate. Some

experts warn that these disrupt hormone function Persuasive research evidence links these ingredients to skin irritations and rashes, dry skin, and even bladder infections.

- Fragrance, which means that the product in question is scented with an artificial chemical or chemicals
- Dyes, such as FD & C Red No. 1, FD & C Blue No. 1 is used to color many popular bath products and shampoos blue.

* Do some research. Authoritative information about hundreds of organic and toxin-free body-care products can be found on the Website of the nonprofit Campaign for Safe Cosmetics (safecosmetics.org).

ORGANIC SOAKS

....

Thanks to the wonders of Mother Nature, you can enjoy a different type of organic soak every day of the week.

Some of the possibilities include:

* Essential-oil baths derived from herbs and trees
  For stimulating the body, relieving congestion, or brightening your mood, just add ten to twelve drops of

an organic essential oil derived from an herb or tree. (Add oil after bathtub is full; swirl around with your hands.)

Try one or more of the following:

Herbs:

- Basil
- Bergamot
- Clary sage
- Lemon balm
- Peppermint
- Rosemary
- Thyme

Trees:

- Cedarwood
- Cypress
- Pine
- Rosewood
- Sandalwood

❀ Essential-oil baths derived from flowers

Comforting, romantic, and really, really relaxing, floral-scented baths calm the heart, quiet the mind, and help lull the body to sleep.

Add ten to twelve drops of an organic essential oil derived from one of the following:

- Geranium
- Chamomile
- Ylang-ylang
- Jasmine
- Melissa
- Lavender
- Rose absolute

* Sea baths incorporating sea salts and seaweed
  Every culture around the world has reveled in the nurturing aspects of seawater, sea salt, and seaweed. For a classic healing sea bath, try adding one to two cups of Dead Sea salts in your tub mixed with ten strips of the Japanese seaweed kombu, which can be found in most health-food stores, many super-

markets, and Asian markets. While the salts tone and clean the body, the kombu detoxifies and conditions skin. If your skin is dry, add a little organic olive oil to your sea bath for extra moisturizing.

Bath Teas Various world cultures have a long tradition of bath teas.

Koreans have been brewing mugwort in tubs for thousands of years. The plant's soothing and sedative properties make it a favorite for nighttime baths; it's also good for helping remedy colds and flus. You can buy organic mugwort in bulk at most health-food stores and from online organic-herb retailers.

In the Aegean and Mediterranean cultures, it is traditional to make a bath tea with lemon or lime flowers and/or lemon, lime, or orange rinds. Tie some up in a bundle and bliss out on the scent of fresh citrus oils.

In the Balearic Islands of Spain, teas are made with

fresh rosemary sprigs, which make for an uplifting bath.

You can turn your bath into a fresh, herbal spa tub by tying up organic herbs in bundles with cheesecloth, muslin, or chlorine-free paper. If you're too busy to make your own, however, health-food stores, beauty-supply stores, and online green body-care retailers often carry premade organic bath teas.

Epsom Salt Soaks For relieving aching muscles and recharging tired bodies, Epsom salts are an excellent, effective, and inexpensive spa treatment. Helping reduce inflammation and swelling associated with bruises, overexertion, sprains, and insect bites, Epsom salts work best when added to the bath while the tap is running. Swirl them around with your hand before climbing into the water; some of them invariably sink to the bottom of the tub.

EPSOM
SALT

Essential-Oil Soaks Derived from Citrus While they offer fragrant relief anytime of year, citrus soaks are particularly clarifying on dark, rainy days or in the dead of winter. Get to the health-food store or visit a purveyor of organic essential oils online to find organic grapefruit, lime, and mandarin orange essential oils.

Organic Soaps and Cleansers The soap or body wash you use depends mainly on your skin type, the season, and whether you are bathing in the shower or tub.

The best-quality organic soaps and body washes are made from such earth-friendly and deep-cleaning ingredients as:

* Saponified organic oils of palm, coconut, olive, and/or palm kernel
* Organic aloe vera juice
* Organic essential oil
* Organic essential-oil blends

- Organic herb, flower, fruit, and marine algae extracts
- Mineral pigment (for color)
- Organic raw shea butter (for moisturizing)
- Organic ground loofah, walnut shells, or apricot pits (for exfoliating)
- Organic honey, powdered milk, or seaweed (for soothing inflamed skin)
- Mud and clay (for clearing oily complexions)
- Filtered spring water

RECYCLING REMINDER

Remember that recycling is an integral aspect to organic bathing. Therefore, please wash out all body-care product containers before you recycle them.

RECIPES FOR FACE, BODY, AND HAIR CARE

....

The art of organic bathing is a full-body pursuit—from the roots of your hair to the tips of your toes. While you can choose from myriad products to indulge yourself naturally, you can also easily, quickly, and inexpensively make your own treatments at home.

....

Organic Face and Neck Masks While you soak, give yourself a facial with prepared products or those you make at home. And don't forget the often-neglected thin and delicate skin on your neck—many times it's sunburned and windburned and needs just as much attention as the skin on your face. On the following pages are some easy and inexpensive recipes.

CLEANSING AND MOISTURIZING MASK FOR DRY SKIN

- 3 medium to large organic avocados
- 1 tablespoon organic avocado oil
- 1 tablespoon organic rose water
- 4 drops organic lavender essential oil

In a large bowl, mash avocados with a fork. Add liquid ingredients one at a time and mix. Gently apply mixture to face and neck. Smooth any extra on décolleté area. Let mask sit 15 minutes. Rinse off with warm water.

BALANCING AND MOISTURIZING MASK FOR NORMAL SKIN

- 1 cup organic rolled oats
- 2½ cups organic milk
- 1 tablespoon organic rose water
- 4 drops organic rose absolute essential oil

To fill your bathroom with the healing aromas of organic essential oils, try one or more of the following. Breathe deeply for maximum benefits.

* Use an aromatherapy diffuser to waft essential oil through the air before, during, and after your bath.
* Sprinkle a few drops of essential oil into your bathwater to scent your bath and help improve moods and physical symptoms.
* Burn organic essential-oil-scented candles as you bathe.

found to enhance hand and eye coordination. Examples include:

- Romantic and sweet: ylang-ylang, jasmine, and rose
- Clean: lavender, iris, geranium, and narcissus
- Tropical: gardenia, pikake, and plumeria

Melissa, chamomile, ylang-ylang, and jasmine all help slow brain waves, elicit relaxation, and ease stress. Research indicates that jasmine activates the limbic system in the brain and stimulates the rod-shaped cells in the retina, on which we rely heavily for night vision. Because jasmine enhances visual tracking and night vision, you may want to use it in your bath ritual when preparing for a night of driving, dancing, or working overtime.

Mixed floral scents of various types have also been

Add rolled oats to milk in saucepan. Bring to boil while stirring. Reduce heat and simmer on low 2 minutes. Oats should look creamy; set aside and let cool 5 minutes. Stir in rose water and essential oil. Apply mixture to face, neck, and décolleté. Keep it on your face for 15 minutes. Rinse off with warm water.

BALANCING MASK FOR OILY SKIN

- 1 large organic cucumber
- 1 cup organic plain yogurt
- 4 drops organic chamomile essential oil

Peel cucumber. Remove the seeds and mash them with a fork in a large bowl. Finely dice remaining cucumber flesh and add to seeds. Stir in yogurt and chamomile oil. Apply in thin layer to face and neck. Let mask sit for 15 minutes, or until it starts to tighten. Rinse off with warm water.

ORGANIC HAIR CARE

....

Conditioning your hair with organic oils while you soak in splendor can make your spa bath a head-to-toe experience.

For normal hair Massage warmed organic olive, almond, or jojoba oil into hair until hair is well coated from roots to ends. Let sit for 30 minutes or more. Shampoo and rinse out oil after the bath.

For dry, permed, or colored hair Massage coconut oil into scalp. Work it through from roots to ends of hair. Shampoo and rinse out oil after the bath.

For oily hair Tea tree oil will help balance an oily scalp and clean up hair. It's also effective against dandruff. Massage into your scalp, then from roots to ends of hair; let sit for 30 minutes. Shampoo and rinse out oil after the bath.

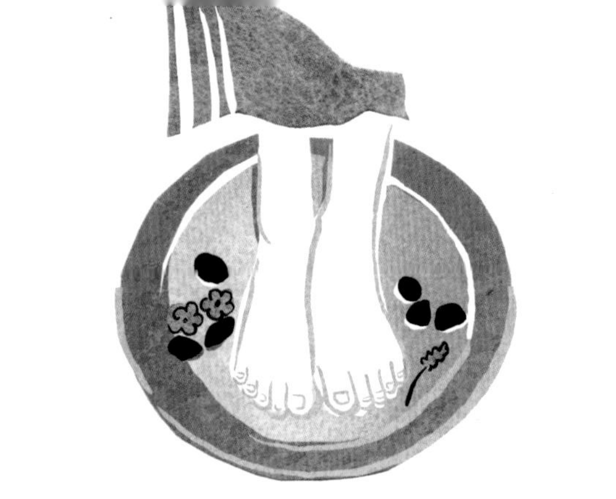

FULFILLING FOOTBATHS

....

A footbath offers quick comfort and valuable benefits for several reasons. Containing thousands of nerve endings, the feet are the foundation of your body and as such exert considerable influence over your mood. When you don't have time for a full-on bath, a footbath can recharge your lower extremities, giving you a much-needed lift.

For soaking feet, use a stainless-steel or ceramic basin.

Footbath tips:

- The classic remedy for tired feet is using one part Epsom salts mixed into two parts hot water.
- For an oceanic foot soak, mix in a few tablespoons of sea salt with the Epsom salts.
- To freshen overheated, tired feet, add 10 to 12 drops of tea tree essential oil after filling the basin with warm to hot water. Swirl it around with your toes. A natural antibacterial and antifungal, tea tree oil helps remedy

athlete's foot. Its astringent properties also help soothe swollen feet and put the spring back in your step.

- To warm up and moisturize winter feet, fill the basin with warm to hot water and add three tablespoons of warmed organic olive oil, four drops of clove essential oil, and a few shakes of powdered cinnamon. The clove oil promotes enhanced circulation and relieves inflammation, while the olive oil heals dry skin.

* After your bath, moisturize feet with a mixture of organic oil (olive, avocado, and coconut are good choices) and a few drops of tea tree oil.

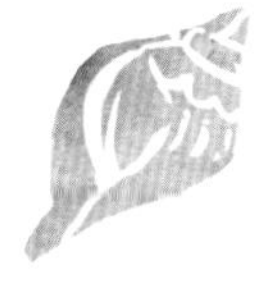

....

Organic Face Care Even if you have oily skin, it's a good idea to moisturize your face, neck, and décolleté after a bath, with an organic oil, body lotion, or cream. This will help you replace skin oils that may have been lost as a result of bathing.

Also remember to rehydrate delicate areas such as around the eyes. Hint: if your skin tends toward oily, use a light organic substance such as jojoba oil.

Organic Body Care Moisturizing after a bath is usually essential (except for those with oily complexions). Most normal, combination, and sensitive skins do well with an after-bath massage of organic olive, avocado, or almond oil. These products are effective as well as inexpensive. Should you prefer an organic cream or lotion, choose the best one you can afford.

Organic Hands-and-Feet Care It's a given that most people will want to moisturize hands and feet after a bath. If you're going to sleep after your soak, follow this simple two-step process to seal in the benefits of the evening's hydrotherapy. Rub rich moisturizers such as organic shea butter or cocoa butter into hands and feet. Slip into organic cotton socks and gloves to maximize moisturizing while you sleep.

AZU's Dreams of Laos
Luang Prabang

Published in 2007 by
AZU Editions (Thailand) Ltd.
111 SKV Building, 8/Fl
Soi Sansabai, Sukhumvit Soi 36
Klongton, Klongtoey
Bangkok 10110
Thailand

Tel: 66 (0)2 712-4016
Fax: 66 (0)2 661-2894
office@azueditions.com
www.azueditions.com

ISBN 978-974-8136-56-1

Printed in Malaysia

Cover: *In the compound of Wat Xieng Thong, one of Luang Prabang's most picturesque temples.*

AZU'S

DREAMS OF LAOS™

Luang Prabang

Photographs by Martin Reeves
Text by John Hoskin

"Where the rivers met was a tiny Manhattan,

but a Manhattan with holy men in yellow robes in its avenues. . . . Down at the lower tip, where Wall Street should have been, was a great congestion of monasteries."

More than half a century has passed since Norman Lewis wrote this captivating description of Luang Prabang, and yet it is scarcely contradicted today. The once royal capital of Laos has endured wars, colonialism, economic depression, and the demise of a monarchy, and yet it seems to resist moving with the times, instead choosing to cling to

Previous spread: *A timeless air prevails in a setting of calm and tranquility.*

Left: *Luang Prabang lies hidden by the folds of high, forested hills.*

its centuries of heritage with the dignity of a dowager duchess in a faded gown.

There may be more foreign visitors these days, but there are still no high-rise buildings in Luang Prabang, no fast-food chains, and few cars. Nor has the population increased that much over the years. It is one of those magical places that immediately charms the traveller, and it appears always to have been so.

In the mid-nineteenth century, the renowned French explorer and naturalist Henri Mouhot undertook a daring expedition along the Mekong River. Upon reaching Luang Prabang, he saw the town as "a charming picture. . . . Were it not for the constant blaze of a tropical sun, or if the midday heat were tempered by a gentle breeze, the place would be a little paradise."

Above left: *There's no need to hurry along Luang Prabang's streets; almost anywhere in the city can be reached on foot.*

Right: *Saffron-robed monks are a characteristic sight in this city of temples and monasteries.*

Hidden amid the timeless folds of the high, forested mountains of northern Laos, the tiny city stands on a promontory where the Mekong River sweeps by on one side and the little Nam Khan tributary flows more gently on the other. It is an extraordinarily evocative spot.

Named after a revered Buddha image that was brought from the temples of Angkor in Cambodia, Luang Prabang was founded in 1353. Although it lost capital status to Vientiane in the sixteenth century, it managed to preserve its standing as the cradle of Lao culture and religion and has remained a treasure house of the nation's finest architectural and artistic achievements.

The superb Wat Xieng Thong, a former royal shrine, and Wat May, distinguished by a magnificent five-tiered roof, are but two of a score of exquisite

Above left: *Luang Prabang's temples are architectural wonders embellished with intricately carved and gilded ornamentation.*

Right: *Decorative detail at Wat XiengThong.*

temples in which masterly architectural lines are complemented by decorative detail that is the very stuff of Oriental fairy tales and legend.

Everywhere a rich design attests to centuries of devotion. Pictured on Wat May's facade, plaster reliefs tell the story of one of the Buddha's former incarnations. Wat Xieng Thong is adorned with gold celestial figures on a black background. Elsewhere murals, woodcarvings, inlay work and other artistic delights lend a dazzling opulence.

Wondrous though the temples are, Luang Prabang is no architectural museum, and its attraction stems essentially from an accumulation of small and often intangible characteristics – the unhurried pace of daily life, the quiet yet warm charm of the people, the scent of wood cooking fires, the sight of saffron-robed monks making their alms round

in the early morning – that are still typical scenes here despite having all but vanished elsewhere.

From the vantage point of Phousi Hill, a pagoda-topped outcrop that dominates the town centre, one looks down on to a view of curving temple roofs and a huddle of simple houses half-hidden by palm fronds and banana leaves. Here you can appreciate life before the package tour and understand why Luang Prabang is regarded as the last original paradise in a fast-developing modern Asia.

Vivid insights are also prompted by the old Royal Palace, which is now a museum. The front rooms contain an interesting collection of religious objects and artworks. Curiously juxtaposed are other exhibits culled largely from the personal possessions of the last monarch, King Sisavang Vatthana. These sundry

Left: *The river, the mountains, and the temples imbue Luang Prabang with a unique, almost mystical, air.*

Above right: *Wat Haw Pra Bang in the grounds of the Royal Palace.*

eft: *Topped by a*
hedi, Phousi Hill
ominates the town
ntre.

ollowing spread:
he view from the
p of Phousi Hill
xtends far beyond
uang Prabang.

items include a desktop stationery set from Thailand, a boomerang from Australia, and a chip of moon rock presented by the United States of America.

Behind the palace's impressive public rooms, however, the private quarters seem spartan by comparison. Like the contrast between art exhibits and personal items, the two styles of apartments offer perspectives on the official and the personal sides of history.

Only Luang Prabang itself seems unaffected by the vagaries of history and the passage of time.

***Above:** Monks take in the view from a vantage point above the Nam Khan River, just outside the town.*

***Right:** Overlooked by Phousi Hill, life in the town proceeds at a leisurely pace.*

***Previous spread:** The city barely peeks out from beneath the jungle canopy.*

***Left:** Novice monks make the climb to the top of Phousi Hill. **Above:** Files of Buddhist monks receive alms in the early morning.*

Above: *Louvred window shutters demonstrate a colonial influence.*

Right: *Colonial buildings make a charming contrast to the religious architecture of the town.*

Above: *Luang Prabang's many surviving old houses are one of the reasons the city is preserved as a World Heritage Site.*

Right: *A quiet stroll along leafy streets is one of the simple pleasures of life in Luang Prabang.*

***Left:** A fruit seller passes along a rather sleepy back street.*

***Above:** Local taxis are as idiosyncratic as Luang Prabang is itself.*

***Above and right:** Kerbside shopping and vendors selling fruits and snacks are an integral part of street life.*

***Following spread:** The night market gears up for business along Luang Prabang's main street.*

30

Left: *A vendor prepares parasols for sale at the night market.*

Above: *Colourful bed quilts are popular buys at the night market.*

Above: *Wat Haw Pra attests to centuries of religious and royal traditon.*

Right: *Luang Prabang's beauty lies in the creations of both mankind and Nature.*

Above: *A Buddhist monk looks out from a window at Wat Xieng Thong.*

Right: *Wat Xieng Thong appears like a scene from an Oriental fairy tale.*

Above: *The imposing five-tiered roof of Wat May is an unmistakable downtown landmark.*

Right: *The defining aspects of Luang Prabang's religious architecture are grace and symmetry.*

Above: *Ornate roof decoration at Wat Haw Pra Bang.*

Right: *A statue of the Buddha depicted with an alms bowl.*

Above and right: *The sacred Pak Ou Caves, where hundreds of Buddha images are enshrined, are reached by boat via the Mekong River.*

Following spread: *The Nam Khan tributary flows into the Mekong at Luang Prabang.*

Left: *A tier of the picturesque, multi-level Kuang Si Waterfall tumbles into a clear turquoise pool.*

Above: *A wide variety of tropical flora is found in the countryside around Luang Prabang.*

Above: *Children at play in the Nam Khan River.*

Above: *After a hard day's work, an elephant enjoys bathtime in the river.*

Left: *An enduring image of fishing on the Mekong River.*

Above: *A fisherman tends his net on the river.*

Following spread: *A picture of sheer serenity as the sun sets over the Mekong.*

Travel Facts

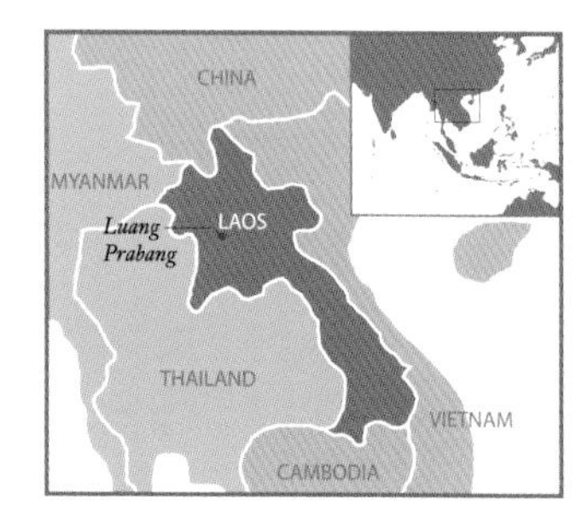

Where It Is

Luang Prabang is situated amid forested rolling hills and mountains in the central part of northern Laos. The city lies on a peninsula at the confluence of the Mekong and Nam Khan rivers. Its geographical coordinates are 19 degrees North, 102 degrees East.

How To Get There

By Air
Luang Prabang International Airport is just north of the city and has links on Lao Airlines to Bangkok and Chiang Mai in Thailand; Hanoi in Vietnam; and Siem Reap in Cambodia. Vietnam Airlines flies from Luang Prabang to Ho Chi Minh City. Lao Airlines flies domestically from Luang Prabang to Vientiane and Pakse. Other destinations connect through Vientiane or Bangkok.

By Road
Luang Prabang is linked to Vientiane and Vang Vieng to the south by Highway 13. The road is mountainous and the journey takes eight to fourteen hours. Buses run to Vientiane several times a day. There are also daily buses to Xiang Khouang and Oudomxai.

By Boat
The Mekong is still an important transport link. Slow boats may travel seasonally between Vientiane and Luang Prabang and between Luang Prabang and Huay Xai upriver on the Thai border. There are also 'speedboats' on the Huay Xai route.

When To Go

Luang Prabang has a tropical monsoon climate with two main seasons: the rainy season, from late May to October, and the dry season, from November to April/May. Temperatures and rainfall can vary depending upon elevation. The average annual temperature is 25 degrees Celsius, with high humidity all year round.

The rainy season sees an average temperature of 32 degrees Celsius and showers on most days. The wettest months are from June to September, when there is also increased humidity.

The hottest months are March, April, and May, when temperatures average 35 degrees Celsius and can reach forty and higher. These months are dry and river levels are low. In the mountains during the coolest months, December and January, temperatures can drop to 15 degrees Celsius and below.

The best time to visit is from November to February, when temperatures are coolest, with the least amount of rainfall.

Main festivals and holidays – when transport and accommodation may be difficult – are Vietnamese Tet and Chinese New Year in January or February; Boun Pimai, the traditional New Year, in April; the Boun Bang Fai rocket festival in May; and the Boun Nam water festival and boat races in October or November.

Find Out More

Comprehensive information for visitors can be found at the Lao National Tourism Administration's website **www.tourismlaos.gov.la**, and also at **www.laoembassy.com** and **www.visit-laos**.

Above: *A Hmong girl smiles for the camera.*

Authors

Martin Reeves *is a British photographer who has been based in Southeast Asia since 1990. He has travelled extensively throughout the region documenting culture, travel, and lifestyle. His work has been frequently exhibited both locally and internationally.*

John Hoskin *is an award-winning freelance travel writer who has been based in Thailand since 1979. He is the author of many highly acclaimed books on travel, art, and culture in Southeast Asia, and has had over 1,000 magazine articles published.*